Physics Problems for GCSE

Alasdair C Shaw

Name	Date

Foreword

Revision is best done actively. Whilst passively re-reading your lesson notes and browsing your revision guide are useful they are unlikely to get you a top mark. One of the best things you can do is answer questions. Lots of them. Once you run out of questions in our textbook and have done all the past papers available you need to look for more problems to solve.

This book contains a wide range of questions aimed at enhancing your abilities at GCSE or iGCSE Physics. Some are simple and repetitive; others are complex and will truly stretch you.
They are grouped into major topics and then further split down into concepts. Not every question will be directly relevant to your specification; if you think you haven't studied something have a quick check in your revision guide to see if you need to know it.

With 300 questions you should certainly be kept busy. Enjoy!

You should also check out A VERY Brief Guide to AQA GCSE Physics at http://lrd.to/z7hT9m3q5k.

If you want revision guides or tests from Alasdair Shaw you could also browse the BBOP: School Physics Resources website at http://www.archaeoroutes.co.uk/edphys...

Forces
and
Motion

Distance and Time

1. The graphs below show how several people got home from school. The axes are not to scale. Match the graphs to the descriptions of the journeys.

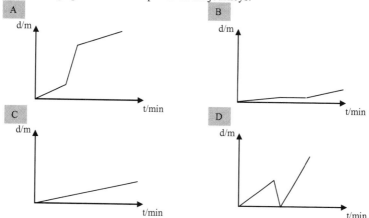

a. Jenny walked home, stopping to buy some crisps on the way.
b. Anja cycled halfway home then remembered something. She pedalled furiously back to school and all the way home.
c. Vincent was collected by his parents. They drove sensibly out of town before turning onto the motorway. After a couple of minutes they left the motorway and drove through their village.
d. Bob ran straight home.

2. A police car joins a motorway on receiving an emergency call and travels north at a constant speed of 30m/s for 5 minutes. It then leaves the motorway at a junction, rejoins it immediately and travels south for 20 minutes at 20m/s to the scene of an accident.
   a. How far did the police car travel in each direction?
   b. How far was the accident from where the police car got the call?

3. A walker travelled a distance of 10km in 2 hours. Calculate the walker's speed in:
   a. km/h
   b. m/s.

4. Calculate the average speed of the following:
   a. a runner who travels 100m in 10s
   b. a cyclist who travels 48km in 3 hours
   c. a car which travels 240km in 3 hours
   d. a rocket which travels 400000km in 80 hours.

5. Prakesh decides to go to Amrit's home to collect some heavy books. The graph shows how his distance from home changes with time.

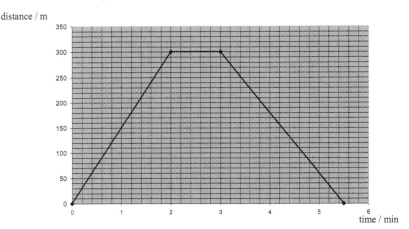

a. How long did it take Prakesh to get to Amrit's house?
b. How far away is Amrit's house?
c. What was Prakesh's average speed on the outward journey?
d. How long did Prakesh stay at Amrit's house?

6. Draw the displacement-time graph for Geoff's journey. He left school on foot and took three minutes to walk 200 metres. He stood waiting for the bus for seven minutes. The bus took him at 1000m/min for 12 kilometres away from school. He got off and walked back towards school at 120m/min, arriving home six minutes later.

7. The distance time graph for a journey is shown on the right.
   Calculate:
   a. the velocity
   b. the acceleration.

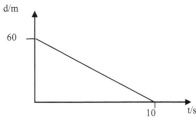

Velocity and Time

8. A cheetah accelerates from rest to 15m/s in 3 seconds. What is its acceleration?

9. An oil tanker can decelerate at $0.002m/s^2$. How long would it take to stop from 10m/s?

10. A train is travelling at 74km/h.
    a. What is its speed in m/s?
    b. How far does it travel in 1 minute?
    c. If the brakes are applied and it decelerates steadily to rest, what is its average speed during the deceleration?
    d. How far does it travel before stopping if it takes 20s to come to a halt?

11. The graph below shows an uncompleted velocity-time graph for someone running a 100m race.
    a. What was the acceleration in the first 4 seconds?

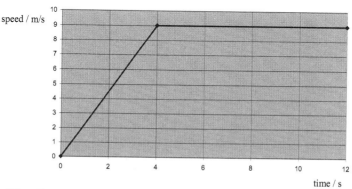

    b. What distance was covered in the first 4 seconds?
    c. Copy and complete the diagram to show that the runner takes 2 seconds to stop after reaching 100m.

12. The distance time graph of another journey is shown on the right.
    Calculate:
    a. the initial velocity
    b. the final velocity
    c. the acceleration in between.

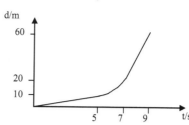

13. In a motor cycle test the speed was recorded at intervals. The readings are shown below.

time / s	0	5	10	15	20	25	30
speed / m/s	0	10	20	30	40	40	40

    a. Plot a speed-time graph of these results.
    b. What was the initial acceleration?
    c. How far did it move in
        i. the first 20s
        ii. the next 10s?

14. A rocket under test reaches a speed of 210m/s from rest in 30s before its fuel is used up. Assuming it accelerates uniformly sketch a speed-time graph of its motion in that time. Use the graph to work out how far it travelled and what the acceleration was before the motor cut out.

15. A car accelerates from rest at 2.5m/s² for 5 seconds. It then cruises for 100m before decelerating to 1m/s over 2s. It comes to rest 3 seconds later.
    a. Sketch the velocity-time graph for this journey.
    b. Sketch the displacement-time graph.

16. A car of mass 1200kg is travelling at a steady speed of 30m/s. The driver brakes and the car stops after 6s.
    Assuming the deceleration is uniform, how fast is the car going 3s after the brakes are applied? Calculate the deceleration.

Scalars and Vectors

17. What is a scalar? What is a vector?

18. Which of the following are vectors?
    speed   velocity   mass   density   acceleration
          force   distance   temperature   depth

19. Explain the difference between speed and velocity.

Forces

20. A deep-space craft executes a 10 second burn of its main rocket motor. When the rocket cuts out the craft is travelling at 100m/s. 10 minutes later the crew prepare to initiate a second burn. What speed is the craft doing at this point?

21. State Newton's Third Law.
    a. A yacht weighs 300 000N. What is the upthrust of the water?
    b. Three people push on a wall, each with 400N. What is the total force of the wall on the people?
    c. A table weighing 900N rests on the floor. What is the force of the floor on each of the four legs?

22. Sketch and label force diagrams for the following situations:
    a. A motionless hot air balloon.
    b. A block accelerating down a frictionless ramp.
    c. A shotgun being fired.
    d. An aeroplane in level flight at constant speed.

23. Calculate the resultant force in each of these cases:
    a. A chair weighing 100N sitting on a concrete floor.
    b. A carriage being pulled by a train with a force of 40000N and against friction of 10000N.
    c. Two gardeners throwing a bag of peat horizontally, one with a force of 300N and the other with a force of 400N.
    d. A horse weighing 5000N, and carrying a rider weighing 800N, jumping upwards with a force of 7000N.

Newton's Second Law

24. What resultant force produces an acceleration of $5m/s^2$ in a car of mass 1000kg?

25. A car of mass 500kg accelerates steadily from rest to 40m/s in 20s.
    a. What is its acceleration?
    b. What resultant force produces this acceleration?
    c. The actual engine force will be greater. Why?

26. What acceleration is produced in a mass of 2kg by a resultant force of 30N?

27. A truck accelerates at $0.5m/s^2$ when acted on by a force of 2200N. What is its mass?

28. An aeroplane of mass 600kg takes off from rest in 50s over a distance of 1500m.
    a. Calculate its speed when it lifts off.
    b. Calculate its acceleration during take-off.
    c. Calculate the force needed to produce this acceleration.
    d. Why is the engine force greater than this in reality?

29. A 1000kg car accelerates from rest to a speed of 10m/s in 20s. Calculate the force required to do this.

30. A lorry of mass 3000kg has a resultant force of 2400N acting forwards on it.
    a. What will happen to the lorry?
    b. Calculate the acceleration.
    c. What force would be needed to decelerate the same lorry at $2m/s^2$?

31. A trailer of mass 1000kg is towed by means of a rope attached to a car moving at a steady speed along a level road. The tension in the rope is 400N.
    a. Why is the tension not zero?
    The car starts to accelerate steadily. The tension in the rope is now 1650N.
    b. With what acceleration is the trailer moving?

32. A train of total mass 30000kg is travelling at a constant speed of 10m/s when its brakes are applied, bringing it to rest with a constant deceleration in 50s.
    a. Sketch a graph to show how the speed of the train changed with time.
    b. Calculate the train's deceleration.
    c. Hence calculate the force of the brakes on the train.
    d. Calculate the ratio of this braking force to the train's weight.

Momentum

33. What is the momentum in each of these cases?
    a. A 0.01kg bullet moving at 250m/s.
    b. A 20kg turkey running at 3.0m/s.
    c. A 0.15kg plate thrown at 10m/s.

34. A 5.0kg ball moving at 3.5m/s hits a stationary 10kg ball and sticks to it. What is the combined velocity after the collision?

Falling

35. A ball is dropped from the top of a tower and takes 1.5 seconds to reach the ground. Which of the following actions will result in a similar ball taking more than 1.5 seconds to reach the ground?
    a. Throwing it downwards from the tower.
    b. Throwing it upwards from the tower.
    c. Throwing it horizontally from the tower.
    d. Dropping it off a taller tower.
    e. Attaching a parachute.
    f. Hollowing out the ball to make it lighter.

36. A 5kg mass is held by a demonstrator using a newton meter. She notes the weight then jumps off a table.
    a. What was the initial weight reading?
    b. What would the weight reading have been while she was in freefall?
    c. Explain the difference.
    d. What would she have felt while she fell?
    e. Was she really weightless?

37. The graph below shows the motion of a skydiver dropping from a balloon.

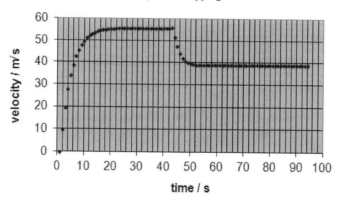

    a. How long after the clock started did he jump?
    b. Describe the motion over the next few seconds.
    c. Explain why the graph flattens off.
    d. What happened next? Why did this have the observed effect on his motion?

38. A frigate fires a gun that drops the spent cartridge case overboard at the same moment the shell leaves the muzzle. The shell gains no aerodynamic lift.
    a. If the barrel is pointing downwards which will hit the ground first, the shell or the cartridge? Why?
    b. If the barrel is pointing horizontally which will hit the ground first? Why?

Stopping Distance

39. The distance taken by a car to stop is made up of two components.
    a. The first part is the thinking distance.
       i. What is the thinking distance?
       ii. List four things that can increase the thinking distance.
    b. The second part is the braking distance.
       i. What is the braking distance?
       ii. List five things that can increase the braking distance.

40. A car at 60mph travels 1 mile every minute. That is 27m every second.
    a. If a driver takes 3 seconds to notice a problem in front of him, how far will he travel during those 3 seconds?
    b. How much less distance would be covered if he only took 1 second to notice the problem?

41. The overall stopping distances for a car travelling at various speeds are listed in the table.

car speed / mph	thinking distance / feet	braking distance / feet	stopping distance / feet
30	30	45	75
40	40	80	120
50	50	125	175
60	60	180	210
70	70	245	315

    a. Using the pattern in thinking distances write down the thinking distance when travelling at 100mph.
    b. What is the pattern relating the braking distance to the speed of the car?
    c. Find the mistake in the stopping distance column. What should it read?

42. The following table gives thinking and braking distances for various speeds.

speed / mph	thinking distance / m	braking distance / m
20	6	6
30	9	14
40	12	24
50	15	38
60	18	55
70	21	75

Plot a graph to show thinking distance, braking distance and stopping distance on the same axes.

43. A car travelling at 30m/s is being driven by a driver whose reaction time is 0.60s.
    a. Calculate the distance travelled by the car in this time.
    b. The maximum possible deceleration of the car is $6.0m/s^2$. Calculate the braking time for the car.
    c. Sketch a speed-time graph for the car, showing the driver's thinking time as well as the braking time.
    d. Show that the stopping distance is 93m.

Moments + Hooke's Law

44. A steel spring stretches by 4cm when it is stretched by a force of 1N applied at one end.
    a. What is the extension of the spring when it is stretched by a force of 5N?
    b. How much force is needed to extend the spring by 10cm?

45. A lorry drives across a beam bridge. The force on each support is recorded in the table below.

distance from support A / m	0	1	2	3	4
force on support A / kN			60		
force on support B / kN		30	60	90	

    a. Copy and complete the table.
    b. What is the weight of the lorry?

46. 500N is placed on one end of a 2m see-saw. What weight must be placed at 0.5m on the other side to balance it?

47. A newton meter is a spring with a calibrated scale.
    a. When a 2N weight is hung on the spring it stretches 5mm. The spring starts to become permanently distorted when 40N are hung on it. Use this information to draw a suitable calibrated scale for this spring.
    b. Indicate on your scale how far the spring would stretch if a 15N weight were hung from it.
    c. The spring stretches 15mm. What weight is hung off it?

48. Which building would be more stable, a tall thin one or a short wide one? Why?

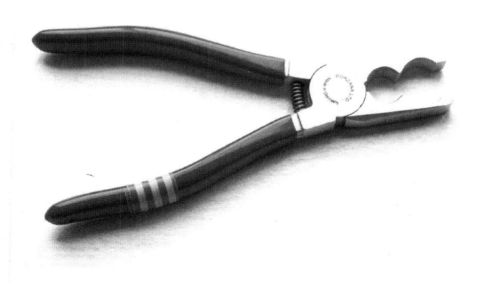

Surface Pressure

49. Calculate the missing values:

force / N	area / m^2	pressure / Pa
20	10	
30		3
	12	4

50. A Sea King helicopter has a weight of 100000N. Its wheels have a total contact area of 0.25m^2 with the ground. What pressure is exerted on the ground?

51. A block measuring 0.1m × 0.4m × 1.5m has a mass of 30kg. Calculate the pressure when each face rests on the ground.

52. Why is walking in bare feet on sand more comfortable than on pebbles?

53. A fire hose is turned on demonstrators. It exerts a pressure of 1000Pa. If the diameter of the jet is 0.6m, what force is exerted?

54. Why are snowshoes used on powder snow? Why are crampons used on ice?

Density

55. Calculate the missing values:

density / kg/m³	mass / kg	volume / m³
	10	5
30		3
4	8	

56. A block of metal has a mass of 2kg. Its dimensions are 0.1m×0.4m×0.1m.
    a. What is its volume?
    b. What is its density?

Fluid Pressure

57. Every ten metres you descend underwater the pressure goes up by one atmosphere. What would be the pressure (in pascals) at the bottom of a 20m deep diver training pool?

58. A tank 4m long, 3m wide and 2m deep is filled with paraffin (density 800kg/m^3).
    a. Calculate the volume of the tank.
    b. Calculate the mass of the paraffin in the tank.
    c. Calculate the weight of this paraffin.
    d. Calculate the pressure on the base of the tank.

59. Describe the behaviour of smoke particles when viewed through a microscope. How does this support the Kinetic Theory of Particles?

60. Explain in terms of particles how gas pressure is caused. Use this to explain the effect on the pressure of halving the volume of a sealed container.

61. A diver at the bottom of a lake releases an air bubble of volume 2cm^3. As the bubble rises its volume increases until at the surface it is 4cm^3.
    a. State the pressure at the surface of the lake.
    b. Calculate the pressure at the bottom.
    c. How deep is the lake?
    Assume the water to be of uniform temperature and that every 10m the bubble rises the pressure drops by 1atm.

62. Air is enclosed in a piston that is kept at room temperature. The end is then pushed so the air now fills only 1/3 the length of the cylinder. What has happened to a) the number of molecules in the air, b) its volume, and c) its pressure?

63. Explain why liquid travels up a straw when you suck on it.

64. Why does water boil at only 50°C on a high mountain?

65. A swimming pool is 3m deep. How much more pressure is there at the bottom than at the surface? (density of water = 1000kg/m^3 gravity = 10N/kg)

Pressure + Temperature

66. Achieving a temperature as close as possible to 0K is the goal of a lot of research.
    a. What does the K stand for?
    b. What is another name given to 0K and why?
    c. What is 0K in °C?
    d. Convert 24°C to K.
    e. Convert 500K to °C.

67. The volume of a sample of dry nitrogen was measured as its temperature changed. The pressure was maintained at 1 atmosphere throughout the experiment. The results obtained are shown in the table below.

temperature / °C	20	50	100	150	200	250
volume / cm^3	48.1	53.0	61.2	69.4	77.6	85.8

    a. Plot a graph of volume (y-axis) against temperature (x-axis). The x-axis scale should begin at -300°C and extend to +300°C with 50°C per division. Draw the best-fit straight line through the points and label it 'line A'.
    b. Use your graph to estimate the volume of the nitrogen at -100°C and 1 atmosphere.
    c. At what temperature would the sample have zero volume?
        i.   What is the significance of this temperature?
        ii. What would happen to the nitrogen gas on cooling before this temperature was reached?

68. In an electrolysis experiment a gas was evolved at an electrode and collected. The volume of gas collected was 60cm^3 at a temperature of 17°C and a pressure of 110kPa.
    a. Convert 17°C into kelvin.
    b. Calculate the volume of this amount of gas at a temperature of 0°C and a pressure of 100kPa.

State Changes

69. Put the three common states of matter in ascending order of energy. Name the fourth state of matter (at a higher energy than the common three).
    For even more show-off points name another two states of matter.

70. Describe the particle behaviour in the three common states of matter.

71. Why does it take energy to melt a substance?

72. Why does it rain when warm damp air is forced upwards?

Energy
and
Power

Energy Transfers

73. Copy and complete the table below.

scenario	energy before	energy after
cooking on a gas oven	chemical	heat and light
shining a torch	chemical	
releasing a clockwork toy		
a parachute falling at a constant speed		
releasing a catapult		kinetic
a TV set showing a film		
a car going uphill at a constant speed		
photosynthesis		

74. Draw transfer diagrams for the following devices:
   a. light bulb
   b. muscle
   c. dynamo
   d. clockwork radio
   e. battery.

75. A competitive cyclist has muscles that are 45% efficient. If 100J of energy are supplied to their muscles, how much goes into kinetic energy?

76. A normal bulb releases 3J of light energy for every 33J of electrical supplied. What is its efficiency?

77. A low-energy bulb has an efficiency of 15%. How much energy is wasted for every 150J of electrical energy supplied?

78. An athlete generates 40J of heat for every 200J supplied to their muscles. How efficient are their muscles if there are no other losses?

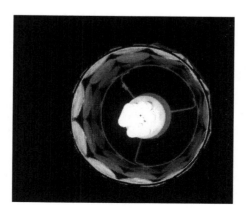

Heat Flow

79. Three beakers are of identical size and shape. One is painted matt black, one is dull white and the third is gloss white. The beakers are filled with hot water.
    a. Which one will cool the slowest?
    b. State a process, other than conduction, convection and radiation, by which heat will be lost from the beakers. How would you reduce this loss?

80. Convection currents can occur in all fluids.
    a. Complete the paragraph below choosing words from the list.
       contract decreases expand fall gases increases rise solids
       "Convection currents can occur in liquids and .... When part of a fluid is heated it expands and its density ...., which causes it to .... Cooling part of a fluid causes it to .... and ...."
    b. Why can you heat water at the top of a test tube without melting ice stuck at the bottom?
    c. Explain why at the coast you tend to get a sea breeze during the day and a land breeze during the night.

81. Dan is cooking the evening meal.
    a. He puts a frying pan on the hob to cook a steak. Explain how the heat conducts through the pan to the meat. Mention particles in your answer.
    b. When he takes the jacket potatoes out of the oven he wraps them in shiny aluminium foil. Explain how this keeps them hot for a long time.
    c. After the meal he boils two kettles of water then leaves them to cool. Both contain 1 litre of water initially at 100°C. One is painted white, the other is painted black. Both cool down at the same rate. What is odd about this? Suggest reasons why this happens.

82. When looking for something to stop heat loss through her attic, Sarah is comparing two products. WonderInsulate has a U-value of $0.2W/m^2$. AmazeUInsulation has a U-value of $0.5W/m^2$. Which is the better insulator?

Heat Capacity

83. A block of copper has a mass of 2.0kg and was heated from 20°C to 50°C . Copper has a specific heat capacity of 385J/kg/°C. How much heat energy was needed to heat it up?

84. A 50kg bath of water gave out 8.4MJ of heat energy as it cooled from 60°C to 20°C. What is the specific heat capacity of water?

Work Done

85. How much work is done in the following situations?
    a. A man pushes a van against a friction force of 300N for 10m.
    b. A mother pushes a pram with a force of 30N for a distance of 100m.
    c. A weight-lifter lifts a weight of 500N through a height of 2m.

86. A gardener pushed a wheelbarrow at a steady speed of 2m/s for 10s using a force of 100N.
    a. How far did she travel?
    b. How much work was done?
    c. Where did the energy come from?

87. An archer pulls back the arrow in his bow a distance of 0.5m against an average force of 200N. How much work was done?

88. A boy with a mass of 60kg climbs 10m vertically up a ladder.
    a. What is his weight?
    b. How much work is done?

89. A weightlifter is in the gym.
    a. On the first machine he lifts 50kg 40 times. Each time he lifts it 30cm. How much work did he do?
    b. On the second machine he does 15 lifts of 80kg. Each lift raises the masses 65cm. How much energy did he use?
    c. He then drops the mass on the second machine down to 60kg. He used 10530J of energy in the next set. How many lifts did he do?

90. A car is travelling along a road with 40 000J of kinetic energy. The brakes are applied and it comes to rest in 20m. Calculate the average braking force.

91. Crash safety is a well-funded area of research.
    a. Explain what a crumple zone is and how it protects people in the car during a crash.
    b. What is special about gases that makes them suitable for airbags?
    c. Why do you think it is illegal not to wear a seatbelt in a moving car?

92. A car has broken down on the isolated winding Cumbrian roads. There's a garage 2000m away along a flat coastal road.
    a. It needs a minimum force of 900N to push the car along a flat road. How much energy would it take to reach the garage by this route?
    There's another garage 1400m away but it's higher up in the fells.
    b. The pushing force up that hill would be 1700N. Calculate the work done if the driver chooses this closer garage.

Gravitational Potential Energy and Kinetic Energy

93. How much work is done when a 3kg mass is lifted vertically through 6m?

94. A hot air balloon and its occupants have a total weight of 5000N. It is descending at a constant speed of 0.5m/s.
    a. What is the upward force on the balloon during the descent?
    b. Calculate its kinetic energy.
    c. Calculate its loss of potential energy in one minute.
    d. Explain why it does not gain kinetic energy.

95. A hiker climbs a hill 300m high. If he has a mass of 50kg calculate the work he does in lifting his body to the top of the hill.

96. A fairground train of total mass 600kg descended a total height of 50m into a dip after it went over the highest point on the track. There was negligible friction and it paused briefly at the top. Calculate:
    a. its loss of potential energy in the descent
    b. its kinetic energy at the bottom of the dip
    c. its maximum speed.

97. A cyclist pedals to the top of a hill, freewheels down the other side gathering speed, applies his brakes and comes to rest at the bottom. Describe the energy transfers taking place.

98. A diver on a high board has 6000J of potential energy.
    a. How much potential energy does she have just before hitting the water?
    b. How much kinetic energy does she have at that point, assuming she is perfectly streamlined?
    c. Where does the energy go when she hits the water?

99. A girl is lifting sandbags from the floor onto a shelf 1.6m high. The pull of the earth on each sandbag is 10N. She lifts 80 bags in 100s.
    a. How much useful work is done in lifting each sandbag?
    b. What is the total potential energy of all the bags on the shelf?
    c. What was her useful power output?
    One sandbag falls off the shelf.
    d. What is its potential energy halfway to the ground?
    e. What is its kinetic energy halfway to the ground?
    f. What happens to its kinetic energy when it hits the ground?

100.   A body of mass 5kg falls from rest and has a kinetic energy of 1000J just before touching the ground. Assuming there is no air resistance and using a value of $10m/s^2$ for the acceleration due to gravity, calculate:
    a. the loss in potential energy during the fall
    b. the height from which the body has fallen.

101.   A tennis ball of mass 0.20kg is released from rest at a height of 2.0m above a concrete floor.
    a. Calculate its potential energy on release.
    b. Calculate its kinetic energy just before impact.
    c. Calculate its speed just before impact.
    After bouncing it rises to a height of 1.5m.
    d. Where does the energy go?

102. An athlete of mass 55kg is capable of running at 10m/s.
   a. Calculate the athlete's kinetic energy at this speed.
   b. If the athlete could convert all this energy into gravitational potential energy, how high could they jump?

Power

103. Copy and complete the following table.

power / W	energy	time
	30J	10s
4000		100s
20	2J	
	6kJ	50s
	5MJ	5min
6000		2hr
100	2kJ	

104. A remote house has a 20kW wind turbine. One day it was able to run for 3 hours to charge a large battery.
    a. How much energy was stored?
    b. For how long could a TV using 500W be run?

105. A boy whose weight is 600N runs up a flight of stairs 10m high in a time of 12s. What is the average power he develops?

106. An elevator is used in a factory to lift packages, each weighing 200N, through a height of 4.0m from the production line to a loading platform. The elevator is designed to deliver three packages per minute to the loading bay.
    a. Calculate:
        i. the potential energy gain of each package
        ii. the time taken for each package to be lifted
        iii. the work done per second by the elevator to lift the packages.
    b. A 200W electric motor is used to drive the elevator. Calculate the efficiency of the elevator and motor system.

107. A passenger aeroplane of mass 25000kg accelerates on a level run from rest to reach its take-off speed of 80m/s in a time of 50s.
    a. Calculate its kinetic energy at take off.
    b. Hence calculate the power developed by its engines.

Resources

108. Solar panels are used to capture the energy from the Sun.
   a. What happens to the solar radiation when it hits the black surface at the back of the panel?
   b. Why is this surface black?
   c. How does the energy get to the oil in the pipes?
   d. Suggest one reason why the pipes are made of copper rather than any other metal.
   e. Why does the oil rise towards the hot water tank?
   f. The hot oil passes its energy to the water in the tank in the 'energy exchanger'. Why is this device at the bottom of the hot water tank?
   g. A lot of energy will be lost from the system.
      i. How will it be lost?
      ii. What can be done to reduce these losses?

109. Why does the National Grid transmit electricity at very high voltage?

110. Discuss the advantages and disadvantages of each of the following:
   a. tidal barrages
   b. nuclear power stations
   c. wind farms
   d. solar cells.

Electricity
and
Magnetism

Plugs and Devices

111. The earth wire conducts current to earth in some circumstances.
    a. What would cause electricity to flow through the earth wire?
    b. Which bit of a plug then cuts off the supply?
    Some devices are coated in plastic.
    c. Why do they only need two wires in their plugs?
    d. What name is given to these devices?

112. Explain how a 5A fuse works.

113. Complete this sketch of a plug, labelling all the parts and indicating the colour of all the wires.

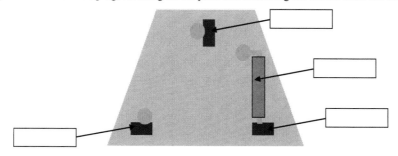

114. An RCD should be placed between the socket and the plug when mowing the lawn.
    a. What does RCD stand for?
    b. Why is it needed when mowing the lawn?
    c. Explain how it works.

115. Sketch a graph of voltage against time for British mains.

Power

116. Define power in terms of energy and time, and then in terms of voltage and current.
    a. How much energy is used in ten seconds by a 60W light bulb?
    b. How much energy is used every second by a 500W motor?
    c. How much energy is used each second when a wire draws a 20A current from a 40V supply?

117. Copy and complete the table below.

power / W	voltage / V	current / A
	12	5
	1.5	0.2
60		10
120	240	

118. A heater has a 13A fuse and runs off the mains. What is its maximum power usage?

119. A small house has six rooms, each with a 60W light bulb. The lights run off the 240V mains. What fuse should be used to protect the lighting circuit?

120. An appliance using 60W blows its 5A fuse. What was the minimum voltage that could have been applied?

121. The manufacturer of a 12V electric heater designed for use in a car claims it has a power of 25W and is capable of heating a flask of tea from 20°C to 50°C in 15 minutes.
    a. Calculate the energy supplied by a 25W heater in 15 minutes.
    b. What current would be drawn by a 12V 25W heater?

Cost

122. A 3kW heater is used for 2 hours.
    a. How many kWh are used?
    b. How many units are used?
    Each unit costs 7p in the daytime.
    c. What was the cost of heating the water during the day?
    d. Why is it better to heat water at night?

123. A 2kW electric heater is used for 10 hours and electricity costs 6p per unit.
    a. How many units (kWh) are used?
    b. What is the total cost?

124. What is the cost of heating a tank of water with a 3000W heater for 80 minutes if electricity costs 7p per unit?

125. An electric cooker has an oven rated at 3kW, a grill rated at 2kW, and two rings each rated at 500W. It operates from the 240V mains.
    a. Would a 30A fuse be suitable for the cooker?
    b. What is the cost of operating all parts for half an hour if electricity costs 6p per unit?

126. A standard light bulb is marked 240V 60W. Peter wants to use as many as possible, with different coloured filters, in his window.
    a. What is the operating current of one bulb?
    b. How many can be run at normal brightness from a 5A fuse?
    c. Draw the circuit that would be used.
    d. What would be the total power of the circuit?
    e. If electricity costs 4p per unit at night, what would be the cost of running the circuit for 5 hours every night for the 12 Days of Christmas?
    f. Give two ways he could reduce the cost.

127. Pat decides to save money by using a pet hamster, Nibbles, to turn a dynamo to make electricity. Nibbles is able to turn his wheel once every two seconds, generating 6J every turn.
    a. What power is Nibbles generating?
    b. What current can be used if Nibbles is to replace a 1.5V battery?
    A new 1.5V battery costs 35p and lasts for 3 hours at this current. Nibbles' food costs 10p for every 10800J he produces.
    c. How long can Nibbles run on 10p of food?
    d. How much would it cost to feed Nibbles to get the same use out of him as a battery?
    e. For how long would Nibbles have to run to save Pat £1?

Series and Parallel Circuits

128. Define a series circuit. Define a parallel circuit.

129. What does an ammeter measure? Explain how to fit one in a circuit.

130. The circuit to the right is set up.
    a. What is component **A**?
    b. What is the power used by the bulb if the current is 5A and the voltage across it is 25V?
    c. If the resistance of **A** were decreased, what would happen to the bulb?
    d. Copy the diagram and add a second bulb in parallel with the first. Is the first or second bulb brighter or are they both the same?

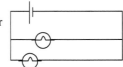

131. In the circuit to the right, what happens to one lamp if the other blows (and so puts a break in the circuit)?

132. Sketch a circuit showing how to measure the voltage across a bulb.

133.

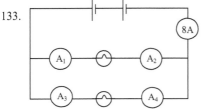

What are the missing ammeter readings?

(You may assume the bulbs to be identical and the wires to be perfect conductors.)

134. In the diagram on the right each cell contributes 3V. What are the missing voltmeter readings?

(You may assume all the bulbs to be identical and the wires to be perfect conductors.)

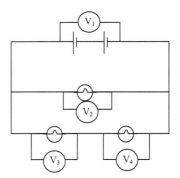

135. A bank has a security alarm. Either the cashier on the front desk or the manager in her office must be able to set off the alarm. They must both also be able to switch it off again if it was a false alarm.

Design a circuit to allow this to be done.

Current-Voltage Characteristics

136. Sketch a graph of current against voltage for a filament lamp. Explain the shape.

137. Sketch the current-voltage graph for a semiconductor diode. Describe the important features. How can a diode be used to control the flow of current in a circuit?

138. Calculate the voltage when:
   a. current = 2A   resistance = 5Ω
   b. current = 9A   resistance = 3Ω

139. Calculate the current when:
   a. voltage = 4V   resistance = 2Ω
   b. voltage = 18V   resistance = 3Ω

140. Calculate the resistance when:
   a. voltage = 6V   current = 2A
   b. voltage = 24V   current = 6A

141. A buzzer requires a current of 7A to sound. It has a resistance of 3Ω. What voltage is required?

142. A battery has a voltage of 1.5V. A bulb connected to it has a resistance of 0.03Ω. What current would flow?

143. A toy train is driven by a 9V battery. What resistance is required to limit the current to 4.5A?

144. How would increasing the resistance in a circuit affect the speed of a motor?

145. A wire has a resistance of 10Ω per metre.
   a. What is the resistance of 50cm of wire?
   b. How would the resistance change as more wire was added?
   c. What would happen to the current?
   d. Suggest a way of making a variable resistor.

Flow of Charge

146. What flows in an electric current through a wire?

147. Draw a series circuit containing a cell, a thermistor and a motor. Indicate the direction of conventional current and of charge flow.

148. What are ions and why do they allow current to flow in a liquid?

149. If the current through a flood lamp is 5A, what charge passes in:
   a. 1s
   b. 10s
   c. 5 minutes?

150. What is the current in a circuit if the charge passing each point is:
   a. 10C in 2s
   b. 20C in 40s
   c. 240C in 2 minutes?

151. A copper coin is attached to the cathode of an electrolysis tank full of copper sulphate solution. It gets coated in copper. What does this tell us about copper and sulphate ions?

152. If a current of 2000A flows through iron sulphate solution, what mass of iron is deposited on the cathode in 2 days?
   (Valency of iron = 3+)
   (Mass of iron ion = 9.3×10^{-26} kg)

153. Light dependent resistors are made of semiconductors and have many uses in controlling circuits.
   a. Describe how increasing the light level affects the resistance of an LDR.
   b. Explain this effect, referring to conduction and valence bands.
   c. Suggest a use of an LDR.

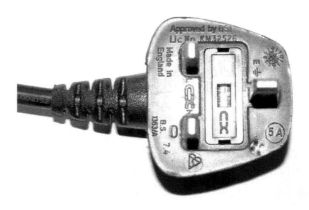

Electrostatics

154. Explain, in terms of electron movement, what happens when a polythene rod becomes negatively charged by being rubbed with a cloth.

155. A car is driven for a couple of hours on a dry day. By the end of the journey the car was negatively charged.
    a. What does negatively charged mean, in terms of electrons and protons?
    b. What would happen if you touched the metal door of the car as you got out?
    c. How could the charge be dangerous to a petrol pump attendant?
    d. How could the charge be used to help repaint the bodywork?

156. When in use the dome of a van der Graaff generator has a positive charge. A small earthed sphere is brought nearby. The positive charge on the dome attracts negative charges from the Earth onto the sphere. The electric force from the dome pulls electrons from air particles creating positive and negative ions.
    a. What is an ion?
    b. Explain how an electric current is conducted between the dome and the small sphere.
    c. What would be observed if the current got very big?

157. Crops are often sprayed with insecticide. Most of the chemical lands on the soil and very little reaches the underside of the plant leaves. To overcome this the spray droplets are given a positive charge as they leave the nozzle.
    a. What happens to the atoms in the spray droplets when they given a positive charge?
    b. What does the charging of the drops do to the pattern of the spray? Explain your answer.
    When the cloud of positively charged drops approach the plants the plants gain a negative charge.
    c. Explain how the movement of ions gives the plants a negative charge.
    d. Explain how this method of spraying crops gives better coverage of the plants and causes less insecticide to be wasted.

158. Aircraft become charged while flying. Explain why aircraft tyres are made using a special rubber that conducts electricity.

Magnets

159. Which of the following statements is the most accurate list of things attracted to a magnet?
   a. plastics and rubbers
   b. metals
   c. iron and steel
   d. iron, steel and carbon
   e. iron, steel, nickel and cobalt

160. Describe two ways to make a permanent magnet.

161. Sketch the shape of the field in each of these cases:
   a.

N   S

   b.

   | N | | S |

   c.

   | N | | N |

162. Electromagnets are very useful.
   a. List three things that affect the magnetic field strength of the electromagnet.
   b. What advantages do electromagnets have over permanent magnets?
   c. Describe one industrial and one medical use of an electromagnet.

163. Sketch the lines of the magnetic field produced by:
   a. a long straight wire carrying a direct current
   b. a solenoid carrying a direct current.

Electromagnetic Induction

164. Alex is given a wire, a pair of magnets and an ammeter.
   She wants to produce an electric current in the ammeter. Explain how she can do this.

165. Explain how a generator works.
   a. State two ways in which the voltage from an alternating current generator would change if its rate of rotation were reduced.
   b. Why is alternating current used to transmit electric power through the national grid system?

166. An aluminium plate placed between the poles of an electromagnet supplied with alternating current becomes warm. Explain how this happens.

Transformers

167. Two coils of wire, A and B, are placed near one another. Coil A is connected to a switch and a battery. Coil B is connected to a centre-zero galvanometer.
   a. If the switch connected to coil A were closed and then opened the galvanometer needle would move. Explain why.
   b. What changes would you expect if a bundle of soft iron wires were placed through the centre of both coils?
   c. What would happen if more turns of wire were wound in coil B?

168. A transformer has 60 turns in its primary coil and 1200 turns in its secondary coil. A 240V 100W lamp is connected to the terminals of its secondary coil. The primary coil is connected to an alternating voltage supply.
   a. Calculate the voltage of this supply if the lamp lights normally.
   b. Calculate the current through the secondary coil when the lamp is on.
   c. Hence calculate the current through the primary coil when the lamp is on.

169. Which of the following statements best describes the main function of a step-up transformer?
   a. Increasing current.
   b. Increasing voltage.
   c. Changing ac to dc.
   d. Changing dc to ac.

170. A transformer is being used to run a 12V bulb off 240V mains. The primary coil has 480 turns. The mains allows a current of 0.5A to pass.
   a. How many turns are there on the secondary coil?
   b. What current will flow in the secondary coil?

171. A transformer has 20 turns on the primary coil and 60 turns on the secondary coil. The supply voltage is 10V and the supply current is 6A.
   a. What is the input power?
   b. If it is a perfect transformer what is the output power?
   c. What is the output voltage?
   d. Use your answers above to work out the output current.

172. A switch mode transformer is one that works at very high frequency.
   a. What are the advantages of using a switch mode transformer over one that works at 50Hz?
   b. What kind of device would this be suitable for?

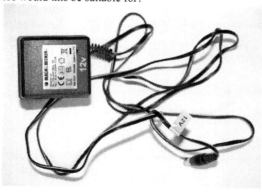

Radioactivity
and
Nuclear

Nuclear Structure

173. An atom of cobalt has an atomic number of 27 and a mass number of 59. Describe simply the structure of the cobalt atom.

174. Natural uranium consists of about 99% $_{92}^{238}U$ and about 1% $_{92}^{235}U$.
    a. How many protons and how many neutrons are present in each type of atom?
    b. What is the name for different types of atoms of the same element?

175. Copy and complete the table below.

name	symbol	what it is	protons	neutrons	electrons
alpha		helium nucleus			
	β				
gamma		photon			

176. What changes, if any, occur in the atomic and mass number of a radioactive nucleus if it emits:
    a. an α particle
    b. a β particle
    c. a γ ray?

177. The following decay chain represents part of a radioactive series. The chemical symbols have been replaced by letters.

$$_{90}^{234}A \rightarrow _{91}^{234}B \rightarrow _{92}^{234}C \rightarrow _{90}^{230}D$$

    a. Name the particles emitted in the three changes.
    b. Write down the two letters that represent isotopes of the same element.

178. Describe the following particle as fully as you can:

$$_{8}^{17}O_2^{3+}$$

Detecting Radiation

179. In a test to identify the type of radioactivity produced by a source the following results were obtained.

material	corrected count rate / Bq
none	450
tin foil	235
1mm aluminium	230
10mm aluminium	228
10mm lead	160

a. Describe the experiment carried out, including a diagram.
b. What does 'corrected count rate' mean?
c. What types of radiation are emitted by this source? Explain your answer.

180. Label the following diagram of a Geiger-Müller tube.

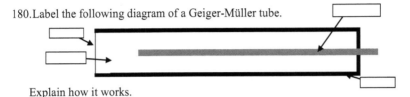

Explain how it works.

181. Radioactivity was first discovered by Becquerel when a lump of uranium ore was left in a drawer with some unexposed photographic film. When the film was developed, on a whim, an image of the rock was revealed.
a. What does this show about the radiation?
b. Use this knowledge to design a badge that could be worn by workers on a nuclear submarine that can be used to find out how much radiation they have been exposed to during a day.

182. A source of ionising radiation is aimed horizontally through a uniform magnetic field. The north pole is on its left. The particles are only detected when the GM tube is placed several degrees above the horizontal. What is being emitted?

183. Background radioactivity accounts for 87% of the exposure to ionising radiations of the average person in Britain.
a. Explain what is meant by:
   i. background radioactivity
   ii. ionising radiation.
b. Describe how you would use a Geiger counter and a stopwatch to measure background radioactivity.

184. A network of satellites was launched during the Cold War to look for the γ-ray discharges from nuclear weapon tests. They started detecting these discharges, but mostly they were not accompanied by ground-based seismograph traces. Eventually the origin of the extra events was traced to outer space. Why is it so hard to build a directional γ-ray detectors? How do you think the direction was discovered?

Measuring Half-Life

185. $^{238}_{92}U$ has a half-life of about 4500 million years. What is meant by half-life?

186. $^{220}_{82}Rn$ emits alpha particles to form an isotope of the element polonium (Po).
  a. Write down the equation for this process.
  b. Radon-220 has a half-life of 52s. A pure sample of this isotope had an initial activity of 400 disintegrations per second. What was the activity after:
    i. 104s
    ii. 208s?

187. Plot the data below and use your graph to determine the half-life of the sample.

time / min	5	7	10	11	14	20	25	27	31	40	50
CCR / Bq	7071	6155	4999	4665	3790	2501	1768	1539	1166	625	312

188. Carbon-14 is an isotope, with a half-life of 5730yrs, naturally occurring in the air. This means that all living things maintain a fixed ratio of carbon-14 to the stable carbon-12. In a 1g sample there would be 180 000 C-14 atoms. This number starts to drop the moment the living thing stops respiring.
  a. A 1g scraping of the charcoal from a cave painting in France contained 45000 C-14 atoms.
    i.   How many times had the number of C-14 atoms halved?
    ii. How many half-lives had passed since the tree (that made the charcoal) was felled?
    iii. When was the painting made?
  b. For many years it was believed that the Romans were the first to introduce grapes to Britain. A 0.02g grape pip found at the British hillfort of Hambledon Hill contained 1800 C-14 atoms.
    i.   How many C-14 atoms would there have been in 1g of grape pip?
    ii. How many half-lives had passed since the grape was picked?
    iii. How long ago was the grape picked?
    iv. What evidence does this give for the introduction of grapes to Britain?

189. The shorter the half-life, the quicker a sample will get used up.
  a. As half-life decreases what happens to the activity (the number of decays per second)?
  b. How would this information affect your choice of radioisotope for use in close proximity to humans?

Uses

190. The thickness of hot rolled steel plate produced in a factory was monitored using a gamma source and detector either side of the sheet after the rollers.
   a. Why was gamma radiation used instead of alpha or beta radiation?
   b. The counter reading increased every second as shown in the table below.

time / s	0	1	2	3	4	5	6	7	8	9
counter reading	0	204	395	602	792	1004	1180	1340	1505	1660

   i.   What was the average count rate over the first 5 seconds?
   ii. What was the average count rate over the last 2 seconds?
   iii. What happened to the thickness of the plate?

191. Cobalt-60 is a radioactive isotope that emits gamma radiation.
   a. Why is it essential to use long-handled tongs to move a radioactive source?
   b. What is gamma radiation?
   c. Why can't alpha or beta radiation be used to treat deep tissue cancer?
   d. With the aid of a diagram explain how the gamma radiation is concentrated on the tumour, minimising the effect on surrounding tissue.

192. A source of radiation is to be injected into a water pipe to check for leaks.
   a. Explain how a leak could be detected.
   b. Use the table below to choose the best isotope for the purpose. Explain your reasoning.

isotope	type of radiation	half-life	soluble
A	γ	2 days	Y
B	β	24 hours	Y
C	β	36 hours	N
D	β	1 month	Y
E	α	36 hours	Y

193. Explain how a suitable isotope could be used as a tracer to determine whether a certain gland is retaining iodine.

Nuclear Models

194. Describe Rutherford's α-particle scattering experiment.
    a. What three types of event were recorded?
    b. What did each suggest about the structure of the gold?
    c. How did Rutherford's nuclear model differ from the plum pudding model?

195. The N-Z curve shows the stable isotopes.
    a. If an isotope appears above the curve on the plot how will it decay?
    b. If an isotope appears below the curve on the plot how will it decay?
    c. If an unstable isotope has N>82 how is it likely to decay?

Fission + Reactors

196. Nuclear reactors and nuclear bombs both rely on chain reactions.
    a. What is a chain reaction?
    b. How is the chain reaction controlled in a nuclear reactor?
    c. What safety feature is built into the design so that the reaction is stopped if there is a power cut?
    d. What material is used as a moderator and what is it for?
    e. Sketch a reactor and add the following labels.
       control rods   fuel rods   moderator   concrete shell   coolant

197. Splitting large nuclei into smaller ones gives off energy.
    a. Describe the process of uranium fission, starting with firing a neutron at a nucleus of uranium.
    b. Write down the nuclear equation for the splitting of a uranium-235 atom into barium-144 and krypton-90. Don't forget the neutrons!

Fusion

198. What is nuclear fusion?

199. Why is nuclear fusion being researched so intently?

200. Nuclear fusion occurs naturally and in man-made devices.
    a. Where does fusion occur naturally?
    b. Write down the nuclear equation for the fusion of two deuterium nuclei into a helium nucleus. (Deuterium is an isotope of hydrogen with an extra neutron.)
    c. Write down the nuclear equation for the fusion of helium and tritium.
    d. What is D_2O commonly called?

Waves
and
Vibrations

Definitions

201. Copy this diagram and add the labels.

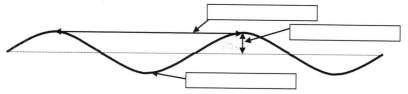

202. Sketch a transverse wave and a longitudinal wave.

203. Copy the diagram below showing wave propagation.

    a. Label a ray and a wavefront.
    b. What do the rays show?
    c. What do the wavefronts represent?
    d. At what angle do wavefronts and rays always meet?

204. A miniature buoy on a pond bobs ten times in one second. The peaks of the ripples passing through are 20cm apart.
    a. What is its frequency of bobbing?
    b. What is the wavelength of the ripples?

205. Which of the following statements are true?
    a. The volume increases as the amplitude of the sound is increased.
    b. The volume increases as the amplitude the sound is decreased.
    c. The volume increases as the frequency the sound is increased.
    d. The note gets higher as the amplitude the sound is decreased.
    e. The note gets higher as the frequency the sound is increased.
    f. It gets darker as the amplitude of the light is decreased.
    g. It gets darker as the frequency of the light is decreased.
    h. It gets redder as the amplitude of the light is increased.
    i. It gets redder as the wavelength of the light is increased.
    j. It gets bluer as the frequency of the light is increased.

206. The aerial of a portable radio is turned through 90°, causing the reception to become weaker. Explain why this happens.

207. A pendulum has a time period of 3.0s.
    a. What is its frequency?
    b. How could the time period be made longer?

208. A piston moves with a frequency of 200Hz. How long does one cycle take?

Reflection

209. A plane mirror is used to produce accurate reflections.
    a. What does 'plane' mean in this context?
    b. What difference is there between a mirror image and a photograph of an object?
    c. If you look in a mirror from 6m away, how far away does your image appear?
    d. If you walk towards a mirror at 1m/s, how quickly does your reflection approach you?

210. Copy and complete these diagrams to show where the light goes.
    a.

    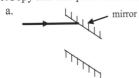

    b.

Refraction

211. Light rays change direction when they pass from one medium to another.
    a. What name is given to this phenomenon?
    b. Which way does a ray bend on passing from air to water?
    c. What property of a wave remains the same either side of the boundary?
    d. Explain why colours look the same in air and clear water.

212. Copy and complete these diagrams to show where the light goes.
    a.

    b.

    c.

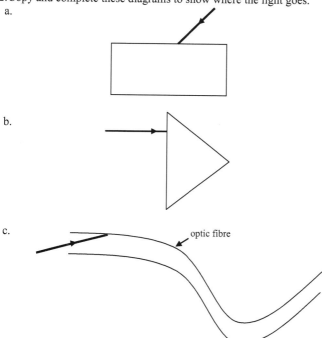

213. Explain what is meant by the 'critical angle' when a ray is shone from perspex to air. State what happens to the ray if its angle of incidence is greater than the critical angle.

214. Describe how optic fibres are used to investigate stomach ulcers.

215. State the Law of Refraction (Snell's Law).

216. A ray of light is incident on a block of glass (refractive index 1.4) at 20°. What is its angle of refraction?

217. A ray of light hits the surface of a pond at 30°. The refractive index of water is 1.34. At what angle does the ray of light enter the water?

Diffraction

218. The design of the breakwaters for harbours is quite complex. In a particular port the gap in the breakwater was 10m across.

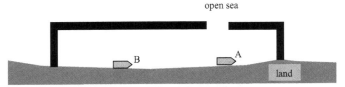

a. In normal wind the waves have a wavelength of 1m. Which of the ships in the diagram will be affected by the waves?
b. In a storm the wavelength approaches 10m. Which ships will be affected now?

219. The diagram on the right represents sound waves travelling from a speaker towards an open doorway. The wavelength of the waves is equal to the width of the doorway.

a. Copy and complete the diagram to show the effects of diffraction as the sound passes through the doorway.

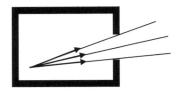

The diagram on the left represents rays of light spreading out from a lamp in the same room.

b. Explain why the light can be treated as rays instead of waves in this situation.
c. Will the light diffract significantly as it passes through the doorway?

d. Hence explain why a conversation can be overheard from outside a room even when it is not possible to see the people.

220. Two-way radios work on a wavelength of several metres. Mobile phones work on a wavelength of a few centimetres. Why do two-way radios often work in valleys when mobile phones don't?

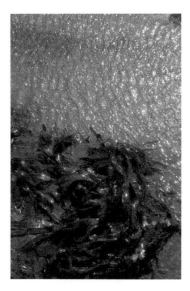

Wave Equation

221. Copy and complete the table below:

wavespeed / m/s	frequency / Hz	wavelength / m
30	10	3
	800	2
	0.25	8
20	5	
0.5	2	
600		50

222. Look at the information below.

a. What is the frequency on which Radio 2 is broadcast?
b. What is the period of the carrier wave for Radio 2?
c. What is the wavelength on which Radio 4 is broadcast?
d. Use these figures to calculate the speed of a radio wave.

223. A 600Hz wave approaches an 11cm gap at 60m/s. Will there be significant diffraction?

224. An ultrasonic cleaning tank operates at a frequency of 40kHz. The speed of sound in water is 1500m/s. Calculate the wavelength of the waves.

225. An industrial microwave, with no turntable, is loaded with a tray covered in butter. After a second in the microwave the tray is removed. There is a regular pattern of melted spots in the butter, corresponding to peaks in the microwave. Each spot is 1.4cm apart. The frequency of the cooker is 10.7GHz.
a. What is the wavelength in cm?
b. What is the wavelength in m?
c. What is the speed of microwaves?
d. Does this seem correct?

226. The speed of sound in air at sea level is 340m/s.
a. Calculate the frequency of sound waves of wavelength 0.10m.
b. Calculate the wavelength of sound waves with a frequency of 5000Hz.

Electromagnetic Spectrum

227. Fill in the gaps in the electromagnetic spectrum shown below.

    .... x-rays .... visible infra-red microwaves ....

228. State a use of each of the following types of wave.
    a. x-rays
    b. radio waves
    c. infra-red
    d. microwaves
    e. visible light.

229. How fast do all electromagnetic waves travel in a vacuum?

230. Different wavelengths of visible light travel at different speeds in glass. Violet light is the slowest.
    a. Which colour of visible light will be refracted the most at an air/glass boundary?
    b. Sketch what would be seen if white light was shone on a glass prism.

231. Electromagnetic waves have different wavelengths.
    a. Which has the longest wavelength?
    b. Which has the highest frequency?
    c. Which of infra-red and ultra-violet has the most energy?
    d. Which of radio and microwaves does the most damage to living tissue?
    e. Which is absorbed by the ozone layer?

232. Describe an experiment to show that heat radiation is mostly concentrated in the infra-red part of the spectrum of light coming from the sun.

233. A soap powder manufacturer decides to mix a substance into the powder which absorbs ultraviolet light and emits visible light as a result. Why would this make clothes seem very white in bright sunlight?

234. Why can an infrared TV camera see animals in darkness?

235. Why is it illegal to coat your car's number plate with a material that is transparent to visible light but reflective to infrared?

Sound

236. The waveform of a note from a musical instrument is shown below.

Copy the diagram and draw on the waveform you would expect if the same instrument were played:

a. more loudly
b. at a lower pitch.

237. Describe how a sound is produced by a guitar, passes through the air, and gets converted into nerve signals.

238. Middle C on a piano is 256. Which of the following does the 256 denote?
a. The length of the piano string in centimetres.
b. The weight of the string in milligrams.
c. The number of vibrations per second.

239. Not all sounds can be heard, just like not all light can be seen.
a. What is the typical range of human hearing?
b. Give two things that cause this range to be reduced in an individual.
c. What name is given to sounds with a frequency too high for a human to hear?
d. Elephants can communicate over long distances using very low frequency sound. What is this called?

240. List four different uses of high frequency sound.

241. A bat is closing in on a moth. It sends out short squeaks of sound and listens for the echoes. The speed of sound in the air is 340m/s.
a. The echo from the moth returns after 0.0012s. How far away is the moth?
b. Another echo returns from a wide area after 0.006s. What is this likely to be? Why is it a lot fainter than the echo from the moth?

242. When using an ultrasonic hospital scanner a paste is applied to the body surface where the probe is used.
a. Explain why this is necessary.
b. Give a reason why the reflected pulse is weaker than the emitted pulse.

Communication

Building Blocks

243. Amplifiers are important parts of most communication systems.
   a. What is the job of a pre-amplifier?
   b. What is the main problem with post-amplifying an analogue signal?
   c. Why don't systems just use a large pre-amplifier instead of a pre-amplifier and a post-amplifier?

244. Explain the function of a modulator.

245. What is the generic term for a device that transforms an input into electrical signals, or vice versa?

246. If a message is to be sent down a wire it could be transmitted by a Morse Code operator or by speaking into a microphone. Both methods used an encoder, a transducer and a transmitter. Give two reasons, in terms of these building blocks, why the second method is faster?

Digital vs Analogue

247. Noise and attenuation are two things that cause loss of signal clarity.
   a. What is noise?
   b. Why is a digital signal better than an analogue signal in a noisy system?
   c. What is attenuation?
   d. Why can boosters be fitted readily to digital systems but not to analogue ones?

248. State whether the following are analogue or digital.
   a. Morse Code
   b. clock with hands
   c. gramophone track
   d. computer signals down a network cable

249. What is the main disadvantage of using a digital system?

Transducers

250. Suggest a transducer for the following conversions:
    a. electrical → sound
    b. electrical → pictures
    c. taps → electrical
    d. sound → electrical
    e. patterns of light → electrical

251. Copy the diagram on the right of a microphone and add these labels:

    cone   coil of wire
    permanent magnet

    How do the sound waves get converted into electrical signals?

Storage

252. Select an appropriate storage system for each of the following situations:
    a. A police 'phone tap that needs to record for many hours. The recording media will be reused if nothing interesting is heard.
    b. A directory of several thousand customers' addresses.
    c. A quick record of a message sent by semaphore.

253. State one advantage of a CD over a magnetic tape and one over a vinyl record.

254. Discuss the advantages and disadvantages of CD-RWs over disks.

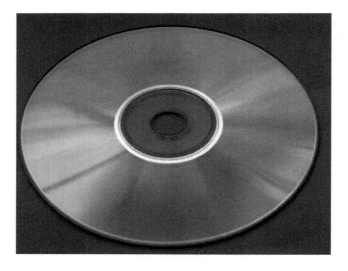

Radio

255. Copy and complete the following paragraph:
Radio is just one part of the .... spectrum. It falls at the .... wavelength end. FM uses VHF, a space wave. This means that it travels in .... Lots of .... are needed to give full coverage of the UK. This means that they have to transmit at slightly different frequencies to avoid ....

256. In a certain part of the country Radio 2 has a carrier wave with a frequency of exactly 90MHz.
a. What is its wavelength?
b. Hence suggest why it can be picked up in some valleys where mobile phones don't work.

257. Give two advantages and two disadvantages of sending Morse code rather than spoken word over a radio link.

258. Major radio stations tend to use FM. Radio hams (amateur radio builders) tend to use AM.
a. What do AM and FM stand for?
b. Sketch an AM wave and an FM wave.
c. Give two advantages of FM radio.
d. Give an advantage of AM radio.

259. A piece of encoded music has a shortest period of 1ms.
   a. Suggest a suitable sample rate.
   b. To what frequency would this correspond?
   c. What wavelength of radio must be used to broadcast this in real time?

260. Explain the qualitative difference between space, sky and ground waves. Use a diagram if necessary.

Satellites + Orbits

261. List 4 uses of passive satellites. List two uses of active satellites.

262. State the one thing that any orbit must cross.
   a. What is the name for an orbit that crosses the Arctic and Antarctic?
   b. What is a geosynchronous orbit?
   c. What name is given for an orbit where the satellite remains above the same point on Earth all the time? Where must that point be?

263. In the following calculations assume the orbit to be circular.
   a. How far would a satellite travel in one orbit of radius r?
   b. If one orbit takes a time T, how fast is the satellite travelling?
   c. How fast must a satellite travel to complete an orbit of radius 10 000km in 16hrs?

264. What is the size of the force keeping a 10kg satellite in a circular orbit of 25000km if it is travelling at 1km/s?

Earth
and
Beyond

Solar System

265. Many things in the sky appear to change at regular intervals. Explain the cause of each of these changes:
    a. The constellations move across the sky every night.
    b. The Moon changes its appearance through a 28 day cycle.
    c. The height of the Sun at midday changes on a yearly cycle.
       On the other hand, some things do not change. The Moon always keeps the same face towards the Earth.
    d. Explain how this is possible.

266. Solar eclipses are very rare.
    a. Sketch a diagram to show how an eclipse of the Sun occurs.
    b. Explain why such an eclipse is so rare.

267. Planets orbit the Sun at different distances.
    a. Why is Mercury very difficult to observe even though it is not too far away?
    b. Why do Pluto and Neptune keep swapping the title of 'furthest planet from the Sun'?
    c. Uranus is closer to the Sun than Neptune. Which takes longer to complete one orbit?

268. What is the name of our sun? Which planet is easiest to see with the naked eye?

269. Describe the formation of our solar system. Account for the Sun, the ecliptic plane, the rotation and orbit of the planets and moons and the differences between the inner and outer planets.

270. Explain how the Moon was formed. What evidence is there for this? Why is there no trace of the event on Earth?

Stellar Evolution

271. Betelgeuse is a red giant. Sol is a small main sequence star.
    a. Why is Betelgeuse described as a red giant?
    b. Describe what will happen to Sol when it uses up all the hydrogen nuclei in its core.
    c. Really massive stars 'die' in a supernova. What is a supernova?

272. Place the terms below in the right order:
      nebula   neutron star   supernova   main sequence   blue supergiant

273. Pulsars give off radio signals.
    a. Why was the first pulsar discovered suspected of being evidence of intelligent extra-terrestrial life, and so labelled LGM1?
    b. What kind of star is a pulsar?

274. Which element is a white dwarf mostly made of?

275. What is a black hole?

Cosmology

276. What is the doppler effect?

277. The Universe is thought to be expanding. Describe the two main pieces of evidence for this.

278. There are several possibilities for the future of the Universe. It may continue to expand, it may reach a steady size, or it may start to contract.
    a. What should cause the expansion to slow down?
    b. What does the evidence currently suggest is happening to the rate of expansion?

279. Hubble noticed that all the galaxies are moving away from ours. The most distant galaxies recede at the greatest rate. Hubble's constant, H_0, describes the rate of expansion of the Universe.
    a. Does this mean we are after all at the centre of the Universe?
    b. Use the data in the following table to plot a graph and determine H_0.

distance / Mpc	20	30	50	60	70	90	110
speed / km/s	1000	1500	2500	3000	3500	4500	5500

    c. Why is knowledge of Hubble's constant important to astronomers?
    d. Suggest why Hubble's constant is difficult to measure accurately.

280. Which of the following is the correct description of 'red shift'?
    1. Red stars are moving faster than blue stars.
    2. Red stars are moving slower than white stars.
    3. The frequency of light detected from a star is lower than the frequency of light that the star emits.
    4. The frequency of light detected from a star is higher than the frequency of light that the star emits.
    5. The light receiver from a star is always closer to red light than when it was emitted.

Medical Uses

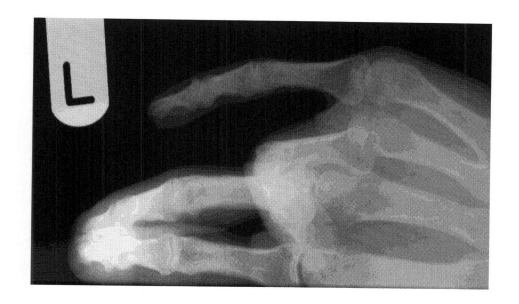

Lenses

281. How does a lens form an image?

282. Draw a ray diagram to show the formation of an image in a convex lens.

283. Does a concave lens form a real or a virtual image? How can you tell?

284. What kind of lens should be used to correct short-sightedness?

285. A magnifying glass is used to look at a 2.0mm long ant. Through the lens it appears 12mm long. What is the magnification of the lens?

286. How large would a 16mm caterpillar look through a lens with a magnification of 5?

287. A lens has a focal length of 2cm. What is its power?

288. What is the benefit of manufacturing a lens from a material of higher refractive index?

The Eye

289. What does the retina do?

290. How do the pupil and iris together react to bright light?

291. How do the ciliary muscles allow the eye to focus on objects at different distances?

292. What two things could cause long-sightedness?

293. A magnifying glass is used to look at a 2.0mm long ant. Through the lens it appears 12mm long. What is the magnification of the lens?

X-rays + Ultrasound

294. Why is ultrasound safer to use than x-rays?

295. What is the advantage of using x-rays compared to ultrasound?

296. An image of x-rays can be obtained using a photographic plate. How else can an image be stored?

297. Ultrasound can be used to detect changes in density by reflection. How is an x-ray image formed?

298. Describe a way that ultrasound can be used to treat a medical condition.

299. Describe a way that x-rays can be used to treat a medical condition.

300. What does a CAT scan provide that a normal x-ray image does not?

About the Author

I studied at the University of Cambridge, leaving with a BA in Natural Sciences and an MSci in Experimental and Theoretical Physics. My masters options included gravitational astrophysics, planetary geophysics, remote sensing and high resolution electron microscopy. I went on to earn a PGCE specialising in Science and Physics from the University of Bangor. A secondary teacher for over ten years I have plenty of experience communicating scientific ideas.

I grew up in Lancashire, within easy reach of the Yorkshire Dales, Pennines, Lake District and Snowdonia. After stints living in Cambridge, North Wales and the Cotswolds I have lived in Somerset since 2002. I have been climbing, mountaineering, caving, kayaking and skiing as long as I can remember. Growing up I spent most of my spare time in the hills.

Landscape archaeology has always been one of my interests; when you spend a long time in the outdoors you start noticing things and wondering how they came to be there. At university I included geophysics in my options.

I am an experienced mountain and cave leader, holding a range of qualifications including ML, SPA and LCL. I am also a course director for climbing and navigation award schemes.

My personal website can be found at http://www.alasdairshaw.co.uk.

To receive email notifications of new physics publications please sign up to my mailing list at http://eepurl.com/bu7HO1.

Also by the same author...

Physics is fundamental to our understanding of the world around us. Everyone should have at least a basic knowledge.

This book explains some of the most interesting parts of physics in simple terms. It dips in, never taking more than two pages for one topic. There is even a good selection of experiments you can do at home!

It is suitable for any student of science as well as anyone wanting to enhance their knowledge. It is a real layman's guide to physics. It treats the reader as an intelligent person who doesn't happen to know about advanced physics; the phrase 'physics for dummies' is often mentioned but that sounds insulting to the reader.

Available from http://www.archaeoroutes.co.uk/bbop/books.php

Some quotes about the book:
"Engages the reader and inspires them to find out more."
"Clear and understandable. I like the conversational style."

"It made physics interesting, even magical."

"This book would be awesome for a high school student or for someone in a career requiring a basic understanding of physics."

Rated 5* by Readers' Favourites

The complete review is available to read at https://readersfavorite.com/book-review/the-best-bits-of-physics.

A few words from the author:

Physics is a hard subject. Everyone knows that. Tell someone that you are studying physics and they will be impressed. This will probably be followed by a comment along the lines of "I could never do that at school". It is still one of the most demanding GCSEs and A-Levels.

This book is an attempt to serve up some of the best bits of physics. It should give you an understanding of the key concepts in modern physics. Along the way I hope you will be convinced that the heart of physics isn't all that hard after all. Perhaps you'll even be able to explain it to your friends and family...

A great present for someone studying physics or anyone who is interested in the world

Physics made easy is one way of looking at it. Physics made interesting is another.

There is now a Teachers' Edition of this book

You can download samples and order it from http://www.archaeoroutes.co.uk/edphys/readers.php. It is packed with A4 worksheets for you to photocopy. Some are comprehension passages and questions; others are experiment instruction sheets.

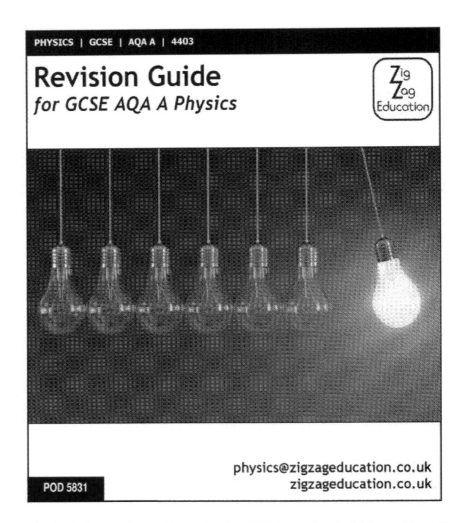

PHYSICS | GCSE | AQA A | 4403

Revision Guide
for GCSE AQA A Physics

Zig Zag Education

physics@zigzageducation.co.uk
zigzageducation.co.uk

POD 5831

You might also be interested in revision guides for GCSE. Currently Alasdair has written guides for AQA A Physics 1 and Physics 2. They are suitable for those studying Separate Physics or Science and Additional Science.

Available from http://www.archaeoroutes.co.uk/edphys/revision.php

A quote about the resources:
"Detailed and comprehensive... The structure is good and leads through the course well. The quizzes are excellent, exam tips are very useful and the exam style questions are very helpful for consolidation and revision"

Sign up for my mailing list to receive notices of new releases: http://eepurl.com/bu7HO1